Fossils Over Time

Q: Are all fossils from dinosaurs?

A: No. There are fossils from plants and animals from long ago. Fossils show what these plants and animals might have looked like.

Leaf fossil

Ocean animal fossils

Q: Were all dinosaurs big?

A: No. There were big dinosaurs
and there were little dinosaurs too.
Some dinosaurs were the same size as you!

This dinosaur is a Troodon (TROH oh don).
Troodons were about 4 feet tall and weighed about
110 pounds. Compare your height to a troodon.

Q: What was the smallest dinosaur?

A: The smallest known dinosaur was the Microraptor (my kro RAP tur). It was 15 inches (39 centimeters) long from nose to tail.

What other things are about 15 inches long?

A beach ball is about 15 inches wide.

Microraptor fossil

Q: What was the largest dinosaur?

A: The Brachiosaurus (brak ee uh SAWR us) was once thought of as the largest dinosaur. It was about 75 feet (23 meters) long. That is about as long as 5 cars!

Triceratops ate plants.

Many dinosaurs were longer than our cars.

How Many Cars Long?		
Brachiosaurus	75 feet	
Stegosaurus	45 feet	
Triceratops	30 feet	
Tyrannosaurus rex	50 feet	

= 15 feet

Q: How tall was a Brachiosaurus?

A: Brachiosaurus was about 50 feet tall. A Brachiosaurus's knee would be above your head. Its leg would be the height of a two-story house.

Second Grader
4 feet
48 inches

Giraffe
16 feet
192 inches

Two-story house
25 feet
300 inches

Brachiosaurus
50 feet
600 inches

Q: How much did a Brachiosaurus weigh?

A: A Brachiosaurus weighed 120,000 pounds.

About 2,000 second graders would equal the weight of a Brachiosaurus! That is about 100 classes of second graders.

Here are some things that weigh as much as a Brachiosaurus.

400 big football players

86 cows

120 grizzly bears

Q: What is the favorite dinosaur?

A: The Tyrannosaurus rex, or T. rex, is probably the favorite dinosaur. It wasn't the biggest. But the T. rex is very well-known.

When we talk about dinosaurs, most people think about the T. rex. Many museums have a T. rex skeleton for people to see.

The brain cavity of a T. rex is big enough to hold only 4 cups of water.

T. rex skeleton

Which dinosaur do your friends like best?
Take a survey.

Use a bar graph. Show your results.

Here is a sample survey.

<table>
<tr><th colspan="2">Our Favorite Dinosaurs</th></tr>
<tr><th>Dinosaur</th><th>Votes</th></tr>
<tr><td>Brachiosaurus</td><td>|||</td></tr>
<tr><td>Microraptor</td><td>||||</td></tr>
<tr><td>Troodon</td><td>卌 |</td></tr>
<tr><td>Tyrannosaurus rex</td><td>卌 ||</td></tr>
</table>

Scientists have found about 700 kinds of dinosaurs. Scientists learn about the dinosaurs by studying fossils.

While you read this book, scientists might be discovering new fossils. There is still a lot to learn!

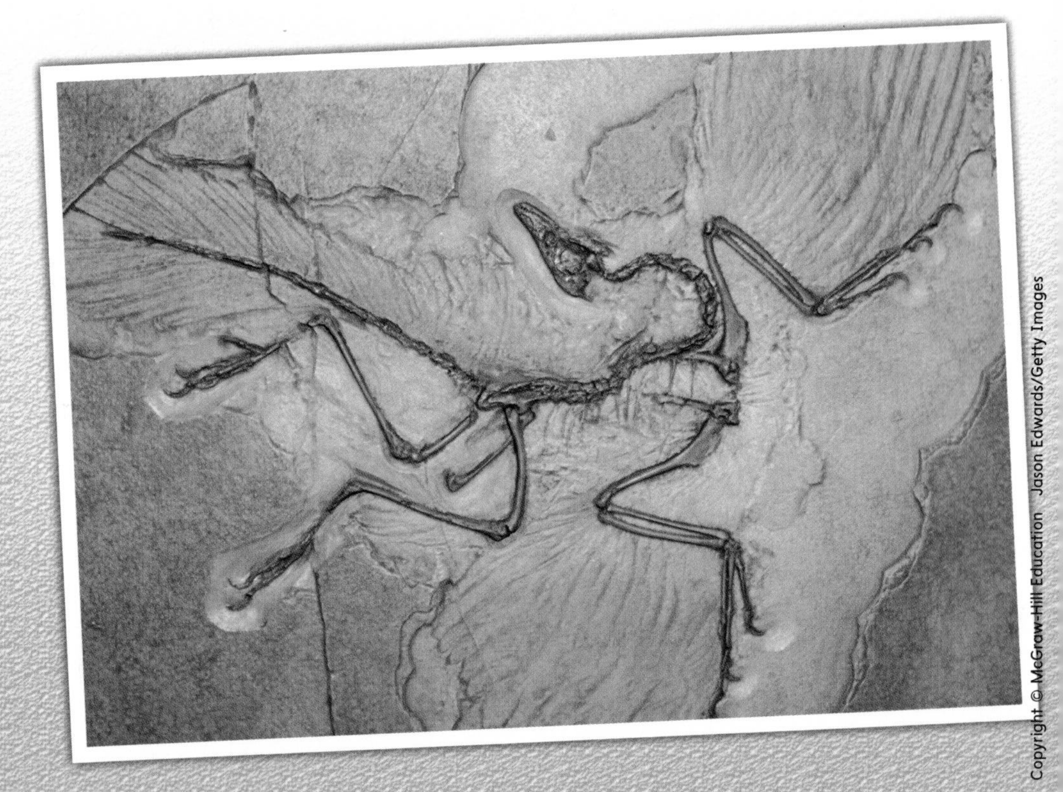